RISING ABOVE THE GATHERING STORM

Developing Regional Innovation Environments

SUMMARY OF A WORKSHOP

Tom Arrison and Steve Olson, *Rapporteurs*

Committee on Science, Engineering, and Public Policy

NATIONAL ACADEMY OF SCIENCES,
NATIONAL ACADEMY OF ENGINEERING, *AND*
INSTITUTE OF MEDICINE
OF THE NATIONAL ACADEMIES

THE NATIONAL ACADEMIES PRESS
Washington, D.C.
www.nap.edu

THE NATIONAL ACADEMIES PRESS 500 Fifth Street, NW Washington, DC 20001

NOTICE: The project that is the subject of this report was approved by the Governing Board of the National Research Council, whose members are drawn from the Councils of the National Academy of Sciences, the National Academy of Engineering, and the Institute of Medicine. The members of the committee responsible for the report were chosen for their special competences and with regard for appropriate balance.

This project was supported by the Wisconsin Alumni Research Foundation, the Morgridge Institute for Research, and the National Research Council. Any opinions, findings, conclusions, or recommendations expressed in this publication are those of the author(s) and do not necessarily reflect the views of the organizations or agencies that provided support for the project.

International Standard Book Number-13: 978-0-309-25604-9
International Standard Book Number-10: 0-309-25604-6

Additional copies of this report are available from the National Academies Press, 500 Fifth Street, NW, Keck 360, Washington, DC 20001; (800) 624-6242 or (202) 334-331; *http://www.nap.edu.*

Printed in the United States of America

THE NATIONAL ACADEMIES

Advisers to the Nation on Science, Engineering, and Medicine

The **National Academy of Sciences** is a private, nonprofit, self-perpetuating society of distinguished scholars engaged in scientific and engineering research, dedicated to the furtherance of science and technology and to their use for the general welfare. Upon the authority of the charter granted to it by the Congress in 1863, the Academy has a mandate that requires it to advise the federal government on scientific and technical matters. Dr. Ralph J. Cicerone is president of the National Academy of Sciences.

The **National Academy of Engineering** was established in 1964, under the charter of the National Academy of Sciences, as a parallel organization of outstanding engineers. It is autonomous in its administration and in the selection of its members, sharing with the National Academy of Sciences the responsibility for advising the federal government. The National Academy of Engineering also sponsors engineering programs aimed at meeting national needs, encourages education and research, and recognizes the superior achievements of engineers. Dr. Charles M. Vest is president of the National Academy of Engineering.

The **Institute of Medicine** was established in 1970 by the National Academy of Sciences to secure the services of eminent members of appropriate professions in the examination of policy matters pertaining to the health of the public. The Institute acts under the responsibility given to the National Academy of Sciences by its congressional charter to be an adviser to the federal government and, upon its own initiative, to identify issues of medical care, research, and education. Dr. Harvey V. Fineberg is president of the Institute of Medicine.

The **National Research Council** was organized by the National Academy of Sciences in 1916 to associate the broad community of science and technology with the Academy's purposes of furthering knowledge and advising the federal government. Functioning in accordance with general policies determined by the Academy, the Council has become the principal operating agency of both the National Academy of Sciences and the National Academy of Engineering in providing services to the government, the public, and the scientific and engineering communities. The Council is administered jointly by both Academies and the Institute of Medicine. Dr. Ralph J. Cicerone and Dr. Charles M. Vest are chair and vice chair, respectively, of the National Research Council.

www.national-academies.org

PLANNING COMMITTEE FOR THE WORKSHOP ON RISING ABOVE THE GATHERING STORM: DEVELOPING REGIONAL INNOVATION ENVIRONMENTS

Judith Kimble (*Chair*), Investigator, Howard Hughes Medical Institute and Professor of Biochemistry and Molecular Biology and Medical Genetics, University of Wisconsin, Madison
Bruce Alberts, Editor-in-Chief, *Science* and Professor Emeritus, Department of Biochemistry and Biophysics, University of California, San Francisco
Ruth A. David, President and Chief Executive Officer, Analytical Services, Inc.
C. D. (Dan) Mote, Jr., Regents Professor and Glenn L. Martin Institute Professor of Engineering, University of Maryland
William J. Spencer, Chairman Emeritus, SEMATECH

Principal Project Staff

Kevin Finneran, Director, Committee on Science, Engineering, and Public Policy
Tom Arrison, Rapporteur and Senior Staff Officer, Policy and Global Affairs Division
Neeraj Gorkhaly, Research Associate, Committee on Science, Engineering, and Public Policy
Steve Olson, Rapporteur and Consultant-Writer

COMMITTEE ON SCIENCE, ENGINEERING, AND PUBLIC POLICY

Richard N. Zare (*Chair*), Marguerite Blake Wilbur Professor in Natural Science, Department of Chemistry, Stanford University
Linda M. Abriola, Dean of the School of Engineering, Tufts University
Claude R. Canizares, Vice President for Research and Associate Provost and Bruno Rossi Professor of Physics, Massachusetts Institute of Technology
Moses H.W. Chan, Evan Pugh Professor of Physics, Pennsylvania State University
Ralph J. Cicerone (ex-officio), President, National Academy of Sciences
Paul Citron, Vice President (Retired), Technology Policy and Academic Relations, Medtronic, Inc.
Ruth A. David, President and Chief Executive Officer, ANSER (Analytic Services, Inc.)
Harvey V. Fineberg (ex-officio), President, Institute of Medicine
Judith Kimble, Investigator, HHMI; Professor of Biochemistry and Molecular Biology and Medical Genetics, University of Wisconsin-Madison
C. D. (Dan) Mote, Jr. (ex-officio), Regents Professor and Glenn L. Martin Institute Professor of Engineering, University of Maryland
Percy A. Pierre, Vice President and Professor Emeritus, Michigan State University
E. Albert Reece, Vice President for Medical Affairs, Bowers Distinguished Professor, and Dean of the School of Medicine, University of Maryland
Susan C. Scrimshaw, President, The Sage Colleges
William J. Spencer, Chairman Emeritus, SEMATECH
Michael S. Turner, Rauner Distinguished Service Professor, Kavli Institute for Cosmological Physics, The University of Chicago
Charles M. Vest (ex-officio), President, National Academy of Engineering
Nancy S. Wexler, Higgins Professor of Neuropsychology, Columbia University

Staff

Kevin Finneran, Director
Neeraj P. Gorkhaly, Research Associate
Marion Ramsey, Administrative Associate

Reviewer Acknowledgment

This report has been reviewed in draft form by individuals chosen for their diverse perspectives and technical expertise, in accordance with procedures approved by the National Academies' Report Review Committee. The purpose of this independent review is to provide candid and critical comments that will assist the institution in making its published report as sound as possible and to ensure that the report meets institutional standards for quality and objectivity. The review comments and draft manuscript remain confidential to protect the integrity of the process.

We wish to thank the following individuals for their review of this report: Jeffrey Alexander, SRI International; David Attis, Education Advisory Board; James Dahlberg, University of Wisconsin; Charles Hasemann, Michigan State University; and Gail McClure, Arkansas Science & Technology Authority.

Although the reviewers listed above have provided many constructive comments and suggestions, they were not asked to endorse the content of the report, nor did they see the final draft before its release. Responsibility for the final content of this report rests entirely with the rapporteurs and the institution.

Contents

1

Overview

In October 2005, the National Academy of Sciences, National Academy of Engineering, and Institute of Medicine released a policy report that served as a call to action. *Rising Above the Gathering Storm: Energizing and Employing America for a Brighter Economic Future* observed that "the scientific and technological building blocks critical to [the United States'] economic leadership are eroding at a time when many other nations are gathering strength." The report laid out 20 recommendations in four broad areas—K-12 education, science and engineering research, higher education, and economic and technology policy—and warned that a failure to take action could have dire economic consequences. As the committee that wrote the *Gathering Storm* report concluded, "we fear the abruptness with which a lead in science and technology can be lost— and the difficulty of recovering a lead once lost."

Rising Above the Gathering Storm sparked intense discussions among policy makers, industrial leaders, and the general public. (See Box 1-1, "The Origins and Aftermath of *Rising Above the Gathering Storm*" at the end of this chapter.) Five years after the release of the *Gathering Storm* report, a second report, *Rising Above the Gathering Storm, Revisited: Rapidly Approaching Category 5*, assessed changes in America's competitive posture. That report concluded that "our nation's outlook has not improved but rather has *worsened*" since the *Gathering Storm* report was released. The report noted examples of other nations that have upgraded their investments in education, technological infrastructure, and innovation systems to a greater extent than has the United States. "In summary," the follow-up report concluded, "the outlook for America to compete for quality jobs has still further deteriorated over the past 5 years."

FOCUSING ATTENTION ON STATES AND REGIONS

The federal government is not the only source of policy actions that can enhance U.S. competitiveness. States and regions within the United States can also contribute to building their capacity for innovation.[1] Areas of intensive innovative activity are scattered throughout the United States—often near major research universities—and all states are interested in strengthening these capabilities.

The ability of the states to drive innovation was the impetus behind a major workshop held in Madison, Wisconsin, on September 20-22, 2011. Entitled "Rising Above the Gathering Storm: Developing Regional Innovation Environments," the workshop brought together leaders in education, government, economic development, and industrial innovation to discuss state and regional initiatives to boost competitiveness through science, technology, and innovation. The conference—which was sponsored by the Wisconsin Alumni Research Foundation, the Morgridge Institute for Research, and the National Research Council—was organized around four major themes:

- Revitalizing K-12 Science and Mathematics Education
- Strengthening Undergraduate Education in Science and Engineering
- Building Effective Partnerships Among Governments, Universities, Companies, and Other Stakeholders
- Fostering Regional Technology Development and Entrepreneurship

The presentations given in each of these four areas are summarized in chapters 2 through 5 of this report. Chapter 6 provides a list of observations and recommendations made by individual participants in the conference's final open-ended discussion. The report has been prepared by the workshop rapporteurs as a factual summary of what occurred at the workshop. The planning committee's role was limited to planning and convening the workshop. The views contained in the report are those of individual workshop participants and do not necessarily represent the views of all workshop participants, the planning committee, or the National Research Council.

AN INNOVATIVE APPROACH

The conference was held at the Wisconsin Institutes for Discovery, which Judith Kimble, Henry Vilas Professor and Howard Hughes Medical

[1] There is an extensive literature on state and regional innovation. One recent collection of insights and perspectives is Charles W. Wessner, Rapporteur; National Research Council. *Growing Innovation Clusters for American Prosperity: Summary of a Symposium*, Washington, DC: National Academies Press, 2011.

Institute investigator at the University of Wisconsin-Madison, described in her opening remarks as the "physical embodiment of the principles of *Rising Above the Gathering Storm.*" (See Box 1-2.) The Wisconsin Institutes for Discovery is a public-private partnership consisting of the private not-for-profit Morgridge Institute for Research and the public Wisconsin Institute for Discovery. Located in a new state-of-the-art facility, it brings together scientists from a broad spectrum of disciplines to conduct research, translate discoveries into applications, enhance cross-disciplinary education, and reach out to the public. As Tashia Morgridge, founding trustee of the Morgridge Institute for Research, said in her opening remarks, the institute has, "as the kids would say, buzz." John Morgridge, chairman emeritus of the Board at Cisco Systems, borrowed a term from his granddaughters to describe the institute: "awesome."

Judith Kimble: "This isn't about making one region strong at the expense of another. We want to make every region in the country strong."

Carl Gulbrandsen, the managing director of the Wisconsin Alumni Research Foundation, said that the great strength of the Wisconsin Institutes for Discovery is the ability "to leverage the human capital and the infrastructure of a great public research university, and . . . leverage the nimbleness and the flexibility of a private research institute." The institution does not have departments, just research themes. The building includes teaching laboratories on each floor, research space, and space for community involvement. "We wanted young people to get excited about science and want to be scientists," said Gulbrandsen. "We wanted people to have fun."

Another great strength of the Wisconsin Institutes for Discovery, said Kimble, is that it is engaged in a positive sum game. Research and education undertaken at the institution make the region stronger while also benefiting the broader society, as the results of research and people trained at the institution move elsewhere. The same can be said of state and regional approaches to innovation in general. Kimble said, "This isn't about making one region strong at the expense of another. We want to make every region in the country strong."

C. D. Mote, Jr.: "The United States has taken actions, but they are too little, they are without long-term commitment, they do not engage those responsible, and they do not reflect an appreciation of the accelerating advancement of other countries."

BOX 1-1
The Origins and Aftermath of *Rising Above the Gathering Storm*

On May 11, 2005, Senator Lamar Alexander delivered a talk entitled "The Next Big Surprise" at the annual meeting of the National Academy of Sciences (NAS). He predicted that within one or two decades other countries would close the economic gap between themselves and the United States. "We need to work together to ensure that our current prosperity is passed on to the next generation," he said.

This talk ignited a "congressional brushfire," said C. D. Mote, Jr., Regents Professor and former President of the University of Maryland, in his remarks at the Madison workshop. On May 27, 2005, the NAS received a bipartisan letter from the Senate requesting responses to specific questions on how to maintain U.S. preeminence in science and technology in the 21st century. On June 30, the NAS received a bipartisan letter containing similar questions from the House of Representatives and requesting a response within 90 days.

Within a few weeks, the NAS, National Academy of Engineering, and Institute of Medicine, through their Committee on Science, Engineering, and Public Policy, formed a 20-member committee that included Nobel laureates, the directors of national laboratories, university presidents, corporate chief executives, and former government officials. Chaired by Norman Augustine, former president and chief executive officer of Lockheed Martin Corporation, the committee met in the summer of 2005 to decide on the top federal policy actions to ensure that the United States would be able to compete, prosper, and be secure in the 21st century.

When the report was released in October of that year, it contained recommendations in four broad areas. The report's highest priority was K-12 science and mathematics education, with a particular focus on the supply of high-quality teachers. The report's second broad recommendation was to support basic research and transformational ideas in science and engineering. The report's third

BOX 1-2
To Learn More

Additional information about the workshop is available at: *http://sites.nationalacademies.org/PGA/COSEPUP/index.htm*

Video of the workshop plenary sessions is available at: *http://vimeo.com/album/1748515*

major recommendation was to attract the best and the brightest into science and technology from both the United States and other countries. And the fourth general recommendation was to create incentives for innovation that would make the United States the premier place in the world to innovate, invest, and create high-paying jobs.

Actions corresponding with many of the report's 20 detailed recommendations were authorized in the America COMPETES Act of 2007. Funds for some provisions of America COMPETES were appropriated in the 2008 supplemental budget. Funding for the Advanced Research Projects Agency-Energy (ARPA-E), which was recommended in the *Gathering Storm* report as a way of undertaking high-risk and potentially high-payoff energy ventures, was appropriated in the American Recovery and Reinvestment Act (ARRA), passed in 2009. Some provisions of America COMPETES have not been funded.

The America COMPETES Act was reauthorized in January 2011. But action on many of the recommendations in the *Gathering Storm* report remains stalled because of constrained resources and political differences in the federal government. "The United States has taken actions," said Mote in his workshop remarks, "but they are too little, they are without long-term commitment, they do not engage those responsible, and they do not reflect an appreciation of the accelerating advancement of other countries. It's fair to conclude that a top priority commitment to U.S. global competitiveness in science and technology is not U.S. policy."

Mote noted that *Rising Above the Gathering Storm* focused on federal actions but that its recommendations extended well past the domain of the federal government. States and localities play major roles in improving K-12 education, accelerating regional economic development, fostering competitiveness within the private sector, and many other issues. "That's why this conference is so important," he said, "because regional and state actions have to be a part of [the solution] if it's going to work."

2

Revitalizing K-12 Science and Mathematics Education

The highest priority actions recommended in *Rising Above the Gathering Storm* were in the area of K-12 education, and three speakers at the Wisconsin conference discussed initiatives in that area. Helen R. Quinn, professor emeritus at the SLAC National Accelerator Laboratory, described a framework for the development of standards in K-12 science education that could greatly improve instruction in science, technology, engineering, and mathematics (STEM). Tom Luce, the founding CEO of the National Math and Science Initiative, discussed two programs with proven track records of success and the prospects for scaling up those and similar programs on a national level. And Michael Lach, special assistant for STEM Education at the U.S. Department of Education, presented some of the initiatives being taken by the Obama administration and described several further steps needed to make progress.

A NEW FRAMEWORK FOR SCIENCE EDUCATION STANDARDS

Based on the success of the common core standards in K-12 math and language arts education, which already have been adopted by many states, the development of standards for K-12 science education was under way at the time of the conference. Helen Quinn described the work of a National Research Council committee that she chaired to develop a framework for those standards.

The goals of the framework were to make possible a coherent investigation of core ideas across multiple years of school and to provide for a more seamless blending of science practices with those ideas and with crosscutting concepts. The committee's report, *A Framework for K-12 Science Education: Practices, Crosscutting Concepts, and Core Ideas*, which was released in July 2011, laid out a vision for K-12 science education and a way to realize the vision.[1] The report specified eight "essential practices for the K-12 science and engineering curriculum." "If you want a definition of 21st-century learning, this is not a bad one," said Quinn. The practices are:

1. Asking questions and defining problems
2. Developing and using models
3. Planning and carrying out investigations
4. Analyzing and interpreting data
5. Using mathematics and information and computer technology
6. Developing explanations and designing solutions
7. Engaging in argument
8. Obtaining, evaluating, and communicating information

"These practices are what scientists do," said Quinn, "and we think students have to do them in order to learn science."

The committee also identified seven crosscutting concepts in science and engineering that students should master:

1. Patterns
2. Cause and effect
3. Scale, proportion, and quantity
4. Systems and system models
5. Energy and matter
6. Structure and function
7. Stability and change

These concepts apply across all fields of science, whether earth systems, biological systems, or physical or chemical systems, and are important in engineering as well. They are the "connective tissue that helps students understand how the pieces fit together," said Quinn.

The document also spells out core ideas for the science disciplines included in K-12 science. The following criteria were used to identify core ideas:

- Have broad importance across multiple science or engineering disciplines, or be a key organizing concept of a single discipline;

[1] National Research Council. *A Framework for K-12 Science Education: Practices, Crosscutting Concepts, and Core Ideas*. Washington, DC: National Academies Press, 2012. Available at: www.nap.edu/catalog.php?record_id=13165.

- Provide a key tool for understanding or investigating more complex ideas and solving problems;
- Relate to the interests and life experiences of students or be connected to societal or personal concerns that require scientific or technical knowledge; and
- Be teachable and learnable over multiple grades at increasing levels of depth and sophistication.[2]

Helen Quinn: "Understanding science and engineering is a tool we use in our lives for making decisions. . . . All students need an understanding of basic science as deeply and as critically as they need to be able to read and do basic arithmetic."

For the physical sciences, the core ideas were (1) matter and its interactions; (2) motion and stability, forces and interactions; (3) energy; and (4) waves and their applications in technologies for information transfer. For the life sciences, they were (1) from molecules to organisms: structures and processes; (2) ecosystems: interactions, energy, and dynamics; (3) heredity: inheritance and variation of traits; and (4) biological evolution: unity and diversity. And for the earth and space sciences, they were (1) Earth's place in the universe, (2) Earth's systems, and (3) Earth and human activity. In addition, the committee specified core ideas in engineering, technology, and the applications of science: (1) engineering design, and (2) links among engineering, technology, and science and society.

The intent is not that these ideas should be tested or taught separately, said Quinn. Rather, curriculum should be designed to help students develop and expand a coherent network of understanding science ideas across multiple years of study. Students should explore a core idea by engaging in scientific and engineering practices and by making connections to crosscutting concepts.

Implementing the Standards

To implement the framework, key components of the K-12 education system need to be aligned, including standards, curricula, instructional materials, assessments, pre-service preparation of teachers, and professional development for in-service teachers, said Quinn. In particular, "having the teachers know where the learning sequence is going and what their job is in the context of that sequence is as important as having the sequence laid out in the curriculum that they are teaching."

[2]Ibid, p. 31.

The framework is designed to produce standards for all students, Quinn emphasized. Every student should have the opportunity to become knowledgable about science and engineering, whether or not he or she decides to pursue a career in science and engineering. "Understanding science and engineering is a tool we use in our lives, both as individuals and as citizens, for making decisions, . . . so all students need to have this basis." The time spent on science in elementary schools has generally decreased as the focus on testing in math and language arts has sharpened, a trend that "is not serving students well," said Quinn. "All students need an understanding of basic science as deeply and as critically as they need to be able to read and do basic arithmetic."

Finally, the standards developed from the framework need to be open to revision. "No set of standards is going to be valid forever," Quinn observed. "We want to know what worked and what didn't work in implementing this vision of science education, so that next time around we'll do it even better." Furthermore, schools and students will continue to change. Education reform needs to be an ongoing and iterative process to adapt to new circumstances and enduring needs.

INCREASING STUDENT ACHIEVEMENT USING PROVEN METHODS

The United States often falls victim to what Tom Luce, founding CEO of the National Math and Science Initiative (NMSI) called pilot disease. "We start a pilot in one school and then we start another pilot in another. And we do pilots very well, but we don't do scale very well."

As a result, NMSI, which was launched in order to implement some of the education recommendations of *Rising Above the Gathering Storm,* uses proven methods to increase student achievement. In this way, it seeks to help all 55 million children in the K-12 public education system, not just 1,000 students here and 1,000 students there.

One program it has promoted is called UTeach, which is training the next generation of K-12 science and math teachers by enabling undergraduates to study the natural sciences and mathematics and simultaneously earn a teaching certificate in those subjects in four years rather than five years. Pioneered at the University of Texas at Austin, UTeach has now been implemented in 33 universities. This replication process is supported by four-year competitive grants of $2.2 million per university awarded by NMSI, and by resource and support materials provided by the UTeach Institute.[3] "Pretty

[3] Additional information about the UTeach program is available at the NMSI Web site (*www.nationalmathandscience.org/programs/uteach-program*) and the UTeach Institute website (*uteach-institute.org/*).

soon we'll be producing 10,000 new teachers a year who are trained in the content," said Luce.

The second program is the Advanced Placement Training and Incentive Program (APTIP), which is designed to provide opportunities for students across the country to take Advanced Placement (AP) math and science courses in high school. When African American or Latino students complete and score a passing grade on an AP course during high school, their college graduation rates go from about 15 percent to more than 60 percent, Luce observed. Furthermore, the AP program is uniform across the United States, enabling high-level college preparation no matter where a student lives.

Pioneered in 10 Dallas schools in 1996, the program has produced "dramatic results," according to Luce. A typical urban high school undertaking the program has 95 percent free and reduced-price lunch enrollment. Yet in the participating Dallas schools, AP passing scores in math, science, and English have increased 11-fold over the past 15 years, and 33 times as many African American and Hispanic students are passing AP tests in those subjects. The program has now been extended to 350 high schools across the country, and the existing teacher corps is being provided with professional development and incentives for completing that training.[4]

Tom Luce: ". . . we do pilots very well, but we don't do scale very well."

"*Rising Above the Gathering Storm* has produced concrete implementation of its recommendation across state lines, across school district lines, across political jurisdictions," said Luce. "I can take you to any state in the union and show you a successful school with any kind of population." The challenge now, he said, is to standardize successful programs like UTeach and APTIP and replicate them across the country. "We don't need another report. We need an implementation plan."

EDUCATION INITIATIVES AT THE DEPARTMENT OF EDUCATION

K-12 education is "incredibly important" to President Obama and his administration, said Michael Lach, special assistant for STEM Education at the U.S. Department of Education. Lach stated that the Obama administration has allocated far more resources to STEM education at the Department

[4] Additional information about the APTIP program is available at the NMSI Web site (*www.nationalmathandscience.org/programs/ap-training-incentive-programs*).

of Education than has any other administration.[5] The President also has expressed his support in less formal ways. He hosted the first astronomy night at the White House and the first annual White House Science Fair. He has invited the winners of math competitions and robotics tournaments to come to the White House, just as standouts in sports and entertainment receive such invitations.

Michael Lach: "We've really pushed to make sure that STEM education is not a standalone piece but is embedded into all of our work."

At the same time, improving STEM education is not just one person's job or the job of the federal government, said Lach. It is everyone's job, including state and local leaders, education administrators, teachers, the business community, and scientists and engineers. "It's going to take all hands on deck working together to make this happen."

Improving STEM education also requires attending to the entire education system, said Lach. Many of the reforms of recent decades have tried to treat science and math as each being in its own silo. But education is too interconnected to treat science and math separately. "It has to be part of how we deal with school funding. It has to be part of how we recruit, hire, train, and contract with teachers. It has to be part of how community groups, museums, and after-school providers all fit into the system. We've really pushed to make sure that STEM education is not a standalone piece but is embedded into all of our work."

The best example of this integration, said Lach, has been the Race to the Top competition. Launched in 2009 with American Recovery and Reinvestment Act (ARRA) funds, the competition has created incentives for states to change the fundamental premises of their education systems.[6] For example, one priority in the competition was for comprehensive statewide plans that focused on STEM. But states were not told to develop a new STEM education system. Rather, the competition directed states to show how and where math and science were embedded throughout their education systems.

---

[5]For an overview of Department of Education K-12 STEM spending, see Executive Office of the President, President's Council of Advisors on Science and Technology. *Prepare and Inspire: K-12 Education in Science, Technology, Engineering, and Math (STEM) for America's Future.* Washington, DC: September 2010, particularly pp. 24-28.

[6] Through the end of 2011, Race to the Top had awarded over $4 billion in grants to 18 states and the District of Columbia. For additional information, see *www2.ed.gov/programs/racetothetop/index.html*.

Fomenting Change

Science and math do have several unique aspects, Lach noted. First, they are sequential, in that they require some concepts to be learned before other concepts. This can be a problem where school districts and states have high mobility rates. When students move from one region to another where math and science education is not coordinated, they can waste time and resources repeating material or trying to learn missed concepts.

Also, many parents are not comfortable with math and science. "It's still okay for someone to go to a cocktail party and make a joke that balancing a checkbook has too much arithmetic involved," Lach said. Given such attitudes, it will take extra effort to teach students the basic concepts described in math and science education standards.

Finally, students, parents, and society in general need to be motivated and inspired to learn math and science. Many students are proficient in math and science but say that they are not interested in the subjects. According to Lach, "a lot of kids have the chops to do this work but, by middle or high school, have been turned off and think it's not for them. So motivation is particularly important."

Education remains fundamentally a state- and local-level activity in the United States, Lach pointed out. The federal government provides just 4 to 5 percent of the total funding for K-12 education, which means that coherent regional plans are incredibly important. Similarly, the help offered K-12 schools from business, higher education, and philanthropy should be coordinated, according to Lach. Colleges and universities also need to focus on the most effective ways to get knowledge to the people who need it, whether teachers, administrators, or parents. "And please, in the standard Calculus 101 class, where the professor gets up and says, 'Look to the left, look to the right—one of you is not going to be here at the end of the semester,' and they say that with all the pride in the world, as if that's a good thing, that kind of culture tells people that [math and science] are only for some, not for all. . . . We have to work together to put that aside."

3

Strengthening Undergraduate Education in Science and Engineering

The challenge of improving U.S. math and science education is not limited to K-12, but instead extends to undergraduate and graduate education. Many of the problems of K-12 education plague undergraduate education as well, as three speakers at the conference explained. Bruce Alberts, professor emeritus at the University of California, San Francisco, and editor-in-chief at *Science* magazine, discussed how to give all students the knowledge and skills they need to be effective workers and citizens in the 21st century. Robin Wright, associate dean in the Department of Genetics, Cell Biology, and Development at the University of Minnesota, provided a concrete example of the kinds of instruction Alberts described. And Lorrie A. Shepard, dean of education at the University of Colorado, Boulder, described some of the cognitive science that supports successful reforms in STEM education.

Bruce Alberts: "If you watch American television today, or listen to our political debates, you have to worry about what's happening to our scientific temper."

THE ROLE OF UNDERGRADUATE EDUCATION IN TRANSFORMING MATH AND SCIENCE EDUCATION

Scientists need to achieve a much higher degree of influence throughout their own nations and the world, Bruce Alberts said. The first Prime Minister of India, Jawaharlal Nehru, called for a "scientific temper" that incorporates the creativity, rationality, openness, and tolerance inherent in science. The mathematician, biologist, and author Jacob Bronowski spoke to the same idea in his 1956 book *Science and Human Values*:

> The society of scientists is simple because it has a directing purpose: to explore the truth. Nevertheless, it has to solve the problem of every society, which is to find a compromise between . . . (the individual and the group). It must encourage the single scientist to be independent, and the body of scientists to be tolerant. From these basic conditions, which form the prime values, there follows step by step a . . . (range) of values: dissent, freedom of thought and speech, justice, honor, human dignity, and self-respect.[1]

Bronowski also wrote of the way in which science has humanized values. As the scientific spirit spread, it generated calls for freedom, justice, and respect. But "if you watch American television today, or listen to our political debates, you have to worry about what's happening to our scientific temper," Alberts said.

Giving Students 21st-Century Skills

The vision of science education laid out in reports like *Rising Above the Gathering Storm* and *A Framework for K-12 Science Education* would prepare children to be problem solvers in the workplace, said Alberts, with the abilities and can-do attitude that are needed to be competitive in the global economy. This vision for science education also precisely fits the needs for workforce skills widely expressed by U.S. businesses. As Ray Marshall and Marc Tucker pointed out in their 1993 book *Thinking for a Living*, the workplace skills needed for success in the modern world economy include:

- A high capacity for abstract, conceptual thinking;
- The ability to apply that capacity for abstract thought to complex real-world problems—including problems that involve the use of scientific and technical knowledge—that are nonstandard, full of ambiguities, and have more than one right answer; and
- The capacity to function effectively in an environment in which communication skills are vital and in groups.[2]

As president of the National Academy of Sciences, Alberts took on the challenge of education reform that would produce these kinds of skills.

---

[1]Bronowski, Jacob. 1956, *Science and Human Values*. New York: Harper and Brothers. pp. 87-88.

[2]Marshall, Ray, and Marc Turner. 1993. *Thinking for a Living: Education and the Wealth of Nations*. New York: Basic Books, p. 80.

"When you look at how the system works, it's the faculties of arts and sciences that define what science education is. They teach the future teachers [and] the parents. What you do in Biology 101 at the University of Wisconsin will define science for those adults for whom that's the last science course they ever have."

However, most science teaching at the undergraduate level does not lead people to understand what science is, Alberts stated. Undergraduates are not allowed to generate and evaluate scientific evidence. They do not learn about the development of scientific knowledge and the difference between scientific knowledge and other forms of knowledge. They do not participate productively in scientific practices and discourse to gain the skills laid out in the framework for K-12 science education.

A major barrier to progress is an overreliance on lecturing, said Alberts. Talking to 500 students in a large lecture hall may be an inexpensive way to teach, but it does not give students the knowledge or skills they need. Many better approaches have been developed that do not entail a great increase in cost, but few faculty members know about them, much less use them.

Alberts' "current obsession" at *Science* is to use the magazine to create more coherence in education. For example, the magazine has named 24 monthly winners of a contest for the best free science education websites.[3] In 2011, it announced a contest for the best inquiry laboratory modules for introductory college science. Modules require between 8 and 50 hours of student work and are readily transferrable so that others can use the same module in their institution.

At the time of the conference, Alberts was working on an editorial about the need for a specially trained scientist in each major school district to connect that district's schools to the wealth of available resources. These specially trained individuals could be connectors or adaptors between the school system and the scientific community. "I have concluded that school districts badly need such an inside person with science in his or her soul who really cares deeply about science and is a scientist." For example, these individuals could coordinate inputs from the local scientific and engineering communities. "I know from my time at UCSF that there are many talented science graduate students and postdocs who would be interested in such a career if a productive new pathway for entry could be developed and promoted."

Efforts to improve science education are going on around the world, Alberts concluded. As noted by the *Gathering Storm* report, the United States needs to take bold steps to meet the competition.

---

[3] Information on the winners is available at *http://www.sciencemag.org/site/special/spore.*

TEACHING MORE BY TALKING LESS

Students do not need more information, said Robin Wright, associate dean, Department of Genetics, Cell Biology and Development at the University of Minnesota. They have a device at their fingertips that provides them with access to all of the knowledge of humankind. What college students need is the ability to do something with that information, to be skeptical, to become sophisticated consumers of information, to create—"the hard part," Wright said.

Robin Wright: "My job now is not the source of all knowledge. It's the coach. I look at my students as collaborators, as emerging colleagues. And they haven't disappointed me once. It's remarkable."

Wright has asked her undergraduates what they should be able to do after four years of college. Among their answers are the following:

- Be ready for more school.
- Apply knowledge, not just have it.
- Think critically and analyze different situations.
- Get a job.
- Develop leadership skills.
- Understand other cultures.
- Manage time well.
- Be able to write well.
- Have experience with research and a laboratory setting.
- Understand scientific writing.

As for what students want from their college instructors, Wright provided the following list:

- Engage us.
- Challenge us.
- Help us develop critical thinking, analytical, and communication skills.
- Make your learning goals transparent to us.
- Provide opportunities for research.
- Use analogies, not jargon.
- Make learning relevant.
- Give us ownership of our learning.
- Infect us with your enthusiasm about the natural world.

Teaching Students to Be Biologists

Wright described an introductory biology course she has helped to develop that encapsulates many of these principles. In the course, students are expected to have read from the textbook and from other resources, and they have a quiz at the beginning of the week that covers that content. Their job in the class is to apply their knowledge, learn how to analyze information, look at data, create new things, and evaluate possible solutions.

Students do not sit in rows. They are in teams at round tables equipped with computers and Internet access. They have to be responsible to and teach each other. By standing in the middle of the room and looking around, the instructor can see how students are doing. "My job now is not the source of all knowledge. It's the coach. I look at my students as collaborators, as emerging colleagues. And they haven't disappointed me once. It's remarkable."

Students take tests first by themselves. They then retake the test in their teams for additional credit. The class average on the test might be 70 percent, said Wright, but it is 95 percent when they take the test as a team. "It's more effective for them to discuss the answers than for us to explain it, in most cases."

During class, students work together on specific problems, such as investigating the structure and function of a molecule. For a take-home exam or paper, they might recommend a strategy to develop a new antibiotic for extremely drug-resistant tuberculosis, or identify a problem of social value and solve that problem using genes. "We are helping them figure out what it's like to think like a biologist," said Wright. "What are the problems biologists wrestle with? What are the resources? What are the limitations?"

Undergraduates have tremendous abilities, said Wright. "They will go to places that are remarkable. If you ever get depressed about all the work ahead of us, all you have to do is think about your students. I'm optimistic about the future."

THE COGNITIVE SCIENCE KNOWLEDGE BASE

Cognitive science research helps to explain why the classes described by Wright work so well, said Lorrie A. Shepard, dean of education at the University of Colorado at Boulder. Active learning does more than keep students awake. Deeper cognitive processes are in play.

A 2002 NRC report contained a list of factors associated with learning:

- Learning with understanding is facilitated when new and existing knowledge is structured around major concepts and principles of the discipline.

- Learners use what they already know to construct new understandings.
- Metacognitive strategies and self-regulatory abilities facilitate learning.
- Learners' motivation to learn and sense of self affect what is learned.
- Participation in social practice is a fundamental form of learning.[4]

Lorrie Shepard: "By engaging students in explaining their reasoning and solving problems collectively, they come away with much higher levels of learning."

Shepard particularly emphasized the last item on this list. People learn language, gestures, interpersonal behaviors, manners, and tastes through interactions with adults and peers. But educators traditionally have had a behavioristic idea of how learning occurs. They advocated memorization first followed by figuring out how to apply that knowledge. Assessments therefore focused on memorization.

Shepard asserted that a fundamental change is needed in how people think about learning. Knowledge makes sense when it is learned in context. Then learners can absorb that knowledge, remember it, and connect it to something they already know. From this perspective, learners need opportunities to interact, explain their reasoning, and use evidence as part of that process.

The National Science Education Standards, which were released in 1996, made this point. But engagement has not been practiced in enough classrooms on a regular basis to become normative, Shepard said. "Scaling up or making these things understood to be common practice is the most difficult task."

Changes in assessments are one of the most effective ways to change teaching, she said. In its 1993 report *Measuring Up: Prototypes for Mathematics Assessment*, the Mathematical Sciences Education Board of the National Research Council listed some of the qualities of mathematical tasks that should be embedded in assessments:

- Promote higher order thinking.
- Draw connections within math, to other subjects, and to life outside of school.
- Emphasize the importance of communicating results.
- Allow for multiple solution strategies.

---

[4]National Research Council. 2002. *Learning and Understanding: Improving Advanced Study of Mathematics and Science in U.S. High Schools.* Washington, DC: National Academy Press.

The success of the mathematics community in improving both instruction and assessments may be a significant reason behind the gradually improving scores of U.S. students on mathematics tests in recent decades, Shepard observed.

The Colorado Learning Assistant Model

Shepard described the Colorado Learning Assistant Model, which embodies the principles she discussed.[5] In this model, outstanding undergraduate students are hired as learning assistants and are given an education course on learning research. The professor lectures twice per week and meets with graduate students and the learning assistants once a week. Instead of "recitations" with graduate students working problems, the learning assistants and graduate students lead "learning teams." Through this process, students are engaged in talking through their reasoning about how to solve problems.

This arrangement creates the kind of interactive environment that is conducive to learning, said Shepard. "By engaging students in explaining their reasoning and solving problems collectively, they come away with much higher levels of learning." Tests of content knowledge show substantial gains in learning after the learning assistant model was instituted. Furthermore, the learning assistants show even greater gains in their mastery of the content—"far above what has been achieved in other contexts." Finally, the learning assistant model also has contributed to a significant increase in the number of undergraduates who are interested in becoming STEM teachers, said Shepard.

The University of Colorado, with support from the American Physical Society, is running workshops across the United States for people who are interested in implementing the learning assistant model. Faculty members need to get over an initial hurdle in learning to use this model. But the success and enthusiasm of the students who benefit from it provide a compelling argument for change.

[5] Additional information is available at *laprogram.colorado.edu/*.

4

Building Effective Partnerships

Rebuilding America's competitiveness is too big a job for any one sector, whether academia, business, or government. The third session of the workshop discussed how partnerships among sectors might also address the challenge.

In this session, three speakers examined the prerequisites for successful partnerships. The director of the Advanced Research Projects Agency-Energy (ARPA-E), Arun Majumdar, reviewed ARPA-E's work and cited the need for alignment of innovation agendas among government, industry, and academia. C. D. Mote, Jr., Regents Professor and former president of the University of Maryland, described the central role of universities in the transition from a national innovation paradigm to a global innovation paradigm. And Mary Good, the Donaghey University Professor at the University of Arkansas, noted the relative lack of a national strategy in the United States compared with other countries and emphasized the potential of universities to step into the breach and support innovation within states and regions.

THE ADVANCED RESEARCH PROJECTS AGENCY-ENERGY

A prominent recommendation of the original 2005 *Rising Above the Gathering Storm* report was for the creation of an agency to foster "out-of-the-box" energy research that industry cannot support due to its high risk. Success in these high risk research areas could produce dramatic benefits for the nation. Arun Majumdar provided a two-year report on the agency and summarized its work as a catalyst of government-university-industry-national lab partnerships.

Majumdar stated that "the mission of ARPA-E is to fund projects that will develop transformational technologies that reduce America's dependence on foreign energy imports, reduce U.S. energy-related emissions, improve energy efficiency across all sectors of the U.S. economy, and ensure that the United States maintains its leadership in developing and deploying advanced energy technologies."[1] The agency's mission is patterned on that of the Defense Advanced Research Projects Agency, which was created during the Cold War to foster radical innovation in defense-related technologies. However, there is a fundamental difference between DARPA and ARPA-E, Majumdar observed. The defense sector is an essentially closed sector of the economy in that it has an eventual customer (the Department of Defense). The energy sector is almost completely open in the sense that the eventual customer could be businesses, consumers, or government. ARPA-E therefore needs to support projects that ultimately will succeed in the marketplace and enable businesses. "It's a different ball game than the DARPA model," he said. ARPA-E looks to create technologies that will induce the private sector to scale up those technologies, "because scaling is what industry does really well."

Examples of Success

Majumdar cited two examples of the approach ARPA-E has taken. The first is the Batteries for Electrical Energy Storage in Transportation (BEEST) program. Instead of incrementally improving the lithium ion battery, the program sought a battery that would give electric cars a longer range and cheaper operating costs than gasoline-based cars. Such a battery needs double the energy density of today's lithium-ion battery at one-third the cost. "This is a really hard problem," said Majumdar.

The program has been supporting several promising approaches. One is an "all-electron battery" being developed at Stanford University that moves electrons rather than ions in the battery. Another is a lithium air battery that provides extremely high energy densities. "People thought this was impossible. Now they're making modules of these, and of course they want to get into manufacturing."

Majumdar also cited a biofuel example. Photosynthesis is an inefficient process for converting sunlight into energy, so ARPA-E is supporting comparable processes with much higher efficiencies. One such process, called Electrofuels, uses microorganisms to harness electrochemical energy from wind, solar, nuclear, or other energy sources to convert carbon dioxide

[1] See *http://arpa-e.energy.gov/About/Mission.aspx.*

into liquid fuels without using petroleum or biomass. Already, teams at the Massachusetts Institute of Technology, the University of Massachusetts Amherst, OPX Biotechnologies, and North Carolina State University have been using such a process to make vials and bottles of oil. "It's not a big plant making millions of gallons, but you've got to start somewhere, and it's a completely new way of making oil."

Arun Majumdar: "Scaling is what industry does really well, and the government does not do well."

Creating New Opportunities to Learn

Rather than following existing learning curves, ARPA-E seeks to create new and much shorter learning curves, said Majumdar. Some of the approaches it is taking will fail. But "we call them not failures, but opportunities to learn, and we go back and try again."

Similarly, rather than relying on a traditional pipeline from research to development to commercialization, ARPA-E tries to compress the process by putting scientists, engineers, technologists and entrepreneurs together. The idea is to blur boundaries so that "scientists educate engineers about what science could do to change the system, and engineers look at systems and educate the scientist about what science is required to change the ball game."

Speed is of the essence, said Majumdar. ARPA-E is asking the technical community to work quickly, and the agency is itself setting records in reviewing proposals and getting funds to investigators. "In the federal government, this is a big deal. Contracting time is now down to two to three months. We're trying to push the system as hard as we can and really engage the stakeholders in this."

Success in the energy field will not occur immediately. It will take at least 15 to 20 years to change energy technologies in the United States or globally in a major way. In the short term, ARPA-E is measuring success in terms of how many good people and projects it is supporting, whether small businesses have been created from universities, how much intellectual property has been created, and the amount of money the private sector is investing in the technologies that are emerging from the agency's efforts.

ARPA-E also measures its success in terms of the partnerships created among government, industry, and academia. The need is for "alignment of innovations," said Majumdar, "not just in science and technology but in finance and markets, policy, education, and society."

GLOBAL INNOVATION PARTNERSHIPS

A transition has taken place where the world has moved from a national innovation paradigm to a global innovation paradigm, said C. D. Mote, Jr., Regents Professor and former president of the University of Maryland. In the past, the federal and state governments were connected to universities and to industry in a national innovation platform. Each sector had its own responsibilities for funding, innovation, education, and commercialization. This partnership contributed substantially to the growth of innovation in the United States and shaped important aspects of that platform.

"This national innovation platform died, basically, around 1990," said Mote, for several reasons. The Cold War ended, which had the effect of adding more than 2 billion people to the knowledge economy. "They wanted to share in the market economy and benefit from it." Also, the Internet was commercialized in 1993 and became the preferred way to communicate globally, with search engines designed specifically to make all information available. High-speed communications spread all over the world, making everyone the neighbor of everyone else. Globalization accelerated through the 1990s.

The Cold War paradigm of "controlling information and innovation" for commercial and security advantages has been replaced by a paradigm of global innovation because information cannot be controlled. Mote cited a sports analogy by describing the Cold War innovation strategy as a defensive strategy. "You don't score as many points as you might, but you keep your opponents from scoring more points than you do, so you win the game. However, if you can't stop your opponents from scoring points, then the only way you're going to win the game is to score more points than they do. That calls for an offensive strategy, and that's the one we're in." Score points fast and often.

The global innovation paradigm is based on partnerships to create information and accelerate innovation. It places an emphasis on talent, employment, markets, manufacturing, research, and investment. It shortens product life cycles, accelerates change, grows consumer markets, and creates a high demand for "front edge" (advanced) employee skills and capabilities. "That's one reason why we have high unemployment and high job opportunities all at the same time," said Mote, because "employers are looking for front-edge skills."

In this new paradigm, partnerships among governments, universities, and industries occur globally on all scales. Governments encompass national, state/provincial, regional, and city governments. Industries are multinational, national, and local. Colleges and universities oversee consortia, campuses, departments, faculty members, and students. "All combinations [occur] on a global platform instead of on a national platform."

C. D. Mote, Jr.: "Universities are ideally positioned to provide innovation services for the global platform enterprise."

Embracing the New Paradigm

"Innovation" has become the answer to almost every "How will we . . . ?" question, said Mote. How will we improve the quality of life for all citizens? How will we stop pandemics? How will we solve climate change? The answer is always "through innovation." One issue is how innovation scales on a global platform. Organizations and individuals tend to scale from the bottom up, while national governments tend to scale from the top down. Regions, communities, and states scale through a hybrid model that combines bottom-up and top-down innovation elements.

Mote emphasized the special role that universities play in this new global innovation paradigm. Universities have almost all of the assets needed for local, regional, national, and global innovation. They can serve as technology incubators, consultants for industry, venture accelerators, conveners of investor networks, and sources of education and training. "Universities are ideally positioned to provide innovation services for the global platform enterprise," said Mote, and "many universities in the United States are working in this direction." One example that he discussed is the University of Maryland-China Research Park.[2]

Because of their unique and irreplaceable assets, universities need to be central players in the global innovation system, Mote said. "In fact, they're the only organizations in our society that can do it."

THINKING GLOBALLY TO ACT LOCALLY

Mary Good, Donaghey University Professor at the University of Arkansas, began by briefly reviewing the work of the Board on Science, Technology, and Economic Policy (STEP) at the National Research Council on international innovation programs. For several years, a STEP committee that she chairs has been comparing innovation systems around the world, including those of China, India, Germany, and smaller countries like Finland. Many governments, she said, have innovation as a national strategic objective. China, for example, has a research park in Shanghai that is led by a woman in San Francisco who recruits companies and people from all over the world to move to the research park to work. Cities like Bangalore

[2] More information on this initiative is available at *www.umcrp.umd.edu/overview.html.*

in India have become fully international, with major companies based in other countries playing prominent roles.

Mary Good: The United States "does not have the focus on strategic development that you see around the rest of the world."

The STEP committee has been particularly interested in manufacturing in Germany, because it combines a large manufacturing base with high wages and social benefits that exceed those of the United States. A particular asset in Germany is the system of Fraunhofer institutes, which are the product of government-university-industry partnerships focused on areas of applied research. These institutes run apprenticeship programs, Good noted, that could be a model for retraining programs in the United States. "We should look at the way those are organized to see if it would not make sense for the United States."

The United States, in contrast to other countries, does not have a national innovation strategy, with just a few exceptions. One exception is the ARPA-E program described by Majumdar, which has the potential to create world-changing industries. Another is the federal Small Business Innovation Research program, though that program "is on life support at the moment," Good noted. Overall, the United States "does not have the focus on strategic development that you see around the rest of the world."

Implications for the States

In contrast, STEP has found several instances of state-level initiatives, often driven by governors, that center on cooperation between government, industry, and universities. State initiatives tend to be different from each other because what works in one state will not necessarily work in another. But those differences are also advantages, because they allow states to take advantage of their unique assets.

New organizations could reignite partnerships similar to those that have worked well in the past. For example, as noted by Mote, some universities are in a position to pull together partnerships, especially with state aid (though states' support for universities has declined markedly, Good observed). State innovation programs also could focus on the medium-size, small, and start-up companies in particular regions, "because in my opinion, that's where most of the jobs in the United States are very likely to get created over the next 15 or 20 years," said Good. These companies are locally situated and can be stable contributors to local partnerships. "People should think about not just partnerships with IBM and General Electric but also with the small [company] across the river that runs a reasonably-sized manufacturing facility and does it well."

Partnerships with universities and governments also can greatly benefit medium-sized and small businesses. Many do not have the resources to secure the modern equipment and analytical technologies needed to remain competitive. Universities, possibly with state support, could be a central laboratory for these kinds of businesses. State manufacturing extension programs also could provide a foundation to build central laboratories.

Finally, Good advocated putting more effort into research parks that involve university-industry-government coalitions and into university incubators. The incubators at the University of Wisconsin and elsewhere could act as models for other universities, "because, frankly, many of us don't run those very well."

The United States needs to figure out how to do things differently than it has in the past, Good concluded. "If we're going to keep our small businesses afloat, we're going to have to figure out how to make these kinds of joint arrangements work."

BOX 4-1
State and Regional Reactions to *Rising Above the Gathering Storm*

Although the original *Gathering Storm* report focused on recommendations for federal policy, the report's message about the importance of education and research to building a strong 21st-century economy resonated with state and regional leaders. Examples of activities focused on state and regional responses to the report include:

- The National Academies organized the Convocation on "Rising Above the Gathering Storm: Energizing and Employing Regions, States, and Cities for a Brighter Economic Future," which was held on September 28, 2006, in Washington, DC. The convocation featured participants from all fifty states.
- At the request of Governor Arnold Schwarzenegger, the California Council on Science and Technology prepared the report *Shaping the Future: California's Response to "Rising Above the Gathering Storm,"* which contained recommendations to the governor and was released in January 2007 (report available at *ccst.us/publications/2006/2006gatheringstorm.php*).
- The Arkansas Science and Technology Authority and the Arkansas STEM Coalition co-sponsored the *Conference on Rising Above the Gathering Storm: Energizing and Employing Arkansans for a Brighter Economic Future,* held September 5, 2007, in Little Rock.
- *The Role of Engaged Universities in Economic Transformation: A Regional Conference inspired by the National Academies Study, "Rising Above the Gathering Storm"* was held October 15-16, 2007, in Ann Arbor, Michigan.

5

Fostering Regional Technology Development and Entrepreneurship

Mary Good commented during a discussion period at the conference: "I don't believe that you're going to get very much help out of the federal government for a while. It's going to be an era where the states and regions are going to lead the federal government, not the other way around." Five speakers provided concrete examples of how the states and regions are doing this. Keynote speaker Duane J. Roth, CEO and member of the board of CONNECT, described how a quarter-century partnership of businesses, universities, and governments transformed the environment for business formation and job creation in San Diego. Sangtae Kim, executive director of the Morgridge Institute for Research in Madison, Wisconsin, explained the rationale behind the institute's creation and pointed toward its immense promise. G. Steven Burrill, CEO of the venture capital firm Burrill & Company, discussed the role of venture capitalists in a radically new entrepreneurial world. Frank Samuel, Jr., President of Geauga Growth Partnership, Inc., provided a specific example of how venture capital has been attracted to the Great Lakes region. And keynote speaker Tommy Thompson, former governor of Wisconsin and secretary of the U.S. Department of Health and Human Services, gave his perspectives on policy making at the state and national levels.

CREATING REGIONAL INNOVATION ENVIRONMENTS

San Diego offers a superb example of what regions can do when they pull together the right ingredients, said Duane Roth, in his keynote address at the conference. CONNECT was founded in 1985 by government, research institutions, and the private sector to aid in the commercialization of research discoveries in and around San Diego. At the time, the region was reeling from economic problems. The imminent end of the Cold War with the Soviet Union was decimating defense contractors, which were major employers in the region. And the savings and loan crisis of the 1980s hit major banks in the region hard.

CONNECT was based on the premise that not nearly enough was being done in the San Diego region to commercialize the excellent research being conducted there. "Nobody in San Diego understood entrepreneurship, venture capital, risk, tech transfer—all the things that have to be done. Until that happens and you educate the community, you're just going to keep publishing papers," said Roth.

Duane Roth: "If you give things, you get more back in return. The collaboration culture is what San Diego became known for."

Spearheaded by leaders at the companies Qualcomm and Hybritech, the University of California, San Diego, and the San Diego Regional Economic Development Corporation, CONNECT has had incredible success over the past quarter century, according to Roth. It has helped start 2,000 companies, has generated about $10 billion in follow-on funding, and has played a role in creating about 150,000 jobs in the region. It now has a staff of 20, membership support of about $3.5 million per year, holds about 350 events annually, and coordinates 2,000 volunteers. Currently, an average of one new company is formed daily in San Diego based on a wide array of enabling technologies.

Formula for Success

CONNECT is based on the formula BC/VC/ED/RC/DC/CC, said Roth, for business creation, venture capital, education, recognition and competitions, Washington, DC, policy, and convergence clusters.

The organization's most important mission is business creation. It does not just arrange for meetings with mentors. Rather, it assigns people and companies an entrepreneur in residence — "a former CEO who is tired of lying on the beach reading a paper and wants to give back." Approximately 300 entrepreneurs in residence currently are helping companies commercialize their ideas.

CONNECT also connects business creators with venture capitalists and trains them in business practices. Minicourses known as frameworks provide a basic background in accounting, finance, intellectual property, and other subjects that entrepreneurs need to know. "These are the kinds of things that we can teach at a very high level, and with this framework scientists and engineers can figure out how those pieces go together."

The organization accords recognition and holds competitions to promote new businesses. It also has an office in Washington, DC, to promote policies that foster rather than discourage business formation. Business leaders come from San Diego to Washington to meet with policy makers and explain their needs. "In Washington, we don't need more voices. We need different voices," said Roth.

Finally, convergence clusters are the means by which many technologies become products. The idea is to encourage companies that are doing similar things to talk with each other. Many companies tend to resist, thinking that they need to keep their ideas private. But in San Diego, companies have learned that "if you give things, you get more back in return," Roth said. "The collaboration culture is what San Diego became known for. It crosses all boundaries."

The idea of collaboration puts the region above individual companies. Companies do not just create jobs and help the economy; they help other companies. Today the San Diego region has more than 250 defense and security companies; more than 3,000 information technology, wireless, and software companies; more than 600 life sciences and biomedical companies; more than 700 energy and environmental companies; and more than 600 sport innovator companies.[1] Together, technology companies employ more than 140,000 people in San Diego, with average tech salaries well above the overall average for the region. Furthermore, new industries such as "information dominance" (defense applications of information technology), wireless sensors in health care, renewable forms of energy, and energy storage have immense promise for the future.

Remaining Challenges

However, said Roth, it is important to recognize that models that have been successful in the past are no longer working today.

First, a new financing model is needed. Fully integrated models in which single companies made discoveries, developed products from those discoveries, and marketed those products are obsolete. As such companies began to focus on production, marketing, and short-term gains, research and innovation often suffered. "Guys like me, product managers, suddenly got into that research room, and we got to talk about the budget. We didn't want anything brand new. We just wanted to make our numbers." Competing divisions within companies tended to undermine support for research

[1] See CONNECT's 2011 Annual Report at *www.connect.org/about/AR2012.pdf*.

and development. And most CEOs were no longer scientists or engineers. As a result, many companies lost their innovation ability.

In addition, passage of the Bayh-Dole Act in 1980 made it possible to transfer publicly funded research into commercial products. This led to many start-up companies that were supported by venture capital and equity sales. But relatively few of these companies succeeded over the long term, and it has been difficult for venture capitalists and the public to pick the successful ones. Instead, large companies have tended to buy promising start-ups for their ideas while slashing jobs, according to Roth.

An alternate approach is to distribute the innovation process, said Roth. Because the outcomes of basic research cannot be predicted in advance, it needs to be funded by the federal government. This process may be inefficient, but "we have to continue to fund [research], because that's where the ideas come from." Product definition companies then could define the product from the discovery to proof of concept. "That's the hole that we have in our system today." By funding midstream development, venture capitalists could fund product development in a portfolio of investments. Furthermore, this work could be very profitable, according to Roth, even if only some of the products being defined end up being commercialized.

In addition, much product definition and development could be "near sourced" by drawing on resources in "the cloud." Building extensive physical infrastructure in startup companies is "terribly inefficient," said Roth. By using resources in the cloud, companies could tap expertise in exactly the areas they need to obtain the data or prototypes to advance their idea. Also, the provision of these cloud resources will be a major new source of employment. This is the approach wireless companies have taken, and they have been extremely successful with it. For example, instead of building their own factories to manufacture phones, they went to factories that already had so much experience and expertise that the wireless companies could not hope to catch up independently. That leaves the companies free to develop the innovations on which future prosperity will be built.

Discussion

In response to a question about manufacturing during the discussion period, Roth described the advantages of manufacturing products in the United States rather than in another country. Modern manufacturing jobs are good jobs that pay well. "We can't let any jobs drift out of this country anymore," he said. Also, manufacturing workers contribute to innovation in many ways by making products better, which is a lesson many other countries have applied in developing their high-technology manufacturing centers.

He also elaborated on the advantages of having the head of the company act as the product manager—someone who understands the science and technology behind products. "Quality gets better, because that CEO who's making the products gets up every day and thinks about quality, lower costs, and not having a product defect or recall."

BRIDGING THE VALLEY OF DEATH

Sangtae Kim: The Morgridge Institute produces a "synergy between open research for the public good and the increased value of companies that live in the ecosystem."

The creation of the Morgridge Institute for Research at the University of Wisconsin-Madison, where the conference was held, was a response to the long and risky pathway from university discoveries to commercial projects. According to Executive Director Sangtae Kim, the Morgridge Institute was designed specifically to bridge the gap between innovation and impact.

The institute focuses on translational research in collaboration with others who have similar interests. It has adopted an approach used by the National Science Foundation in establishing the NSF Engineering Research Centers, which involves three layers of knowledge. The foundation layer consists of the scientific discoveries emerging from university research. The integrative layer relies on a human, technological, and physical infrastructure to produce an innovation environment. And the summit layer involves the delivery of solutions to overarching problems.

Kim said that he also draws on his experience at the Parke-Davis Research Center in the 1990s. Though the budget of the center was small compared with other pharmaceutical R&D centers, it created some of the most successful pharmaceuticals of the time, including Lipitor. The success of the center, said Kim, hinged on the role of project managers. Led by CEO Ronald Cresswell, the center made project managers into drug development vice presidents with rank, budget authority, and power equal to that of vice presidents in functional areas of the company, such as toxicology or chemistry. As a product like Lipitor moved from discovery biology through optimization, toxicology, clinical studies, and eventually to sales and marketing, the drug development vice president followed that product through the functional departments of the company. "Very powerful project management teams were superimposed on top of strong functional areas and capabilities," said Kim.

In universities, individual professors are analogous to the vice presidents of functional areas. They are rewarded for continued excellence in their area of expertise. When they publish a paper, they typically do not tend to pursue the idea into domains where they lack expertise. Their instinct is to return to their areas of expertise and try to replicate their success. Meanwhile, the project managers in a university setting are typically associated with the technology transfer office and have to scramble to link discovery with the delivery of innovations. "It's not surprising that innovation from university campuses to product is a haphazard situation," said Kim.

The Morgridge Institute seeks to achieve a balance between discovery and delivery, said Kim. This process is complicated by the fact that initial ideas for innovation often do not work out. As a result, investors often lose patience before an idea makes it across the "valley of death" to commercialization.

The Morgridge Institute helps ideas through the valley of death in part by working with local consortia that together represent a critical mass of capabilities in the region. The institute can support research on alternative pathways and better ways of achieving an end while companies focus on the development of products and avoid diverting their capital to precompetitive research. The result is a "synergy between open research for the public good and the increased value of companies that live in the ecosystem."

Kim cited an example called the Accelerators in Medicine consortium, which is focused on the use of particle accelerators in medicine. No one company has the resources to build instruments that can be used for such purposes as proton therapy for tumors, miniaturized mass spectrometers, and solid state CT scanners. Instead, a half dozen companies, the Morgridge Institute, and the University of Wisconsin-Madison have pooled their resources in an R&D consortium. The consortium reduces the technical risk and makes it easier for the companies to attract investment capital. One medical isotope company, for instance, was able to raise $20 million and significantly ramp up its scale of operation after working with the consortium for 18 months. Furthermore, said Kim, the value of the royalty streams to the Morgridge Institute from the sale of the medical isotope, if the project works, would be in the many hundreds of millions of dollars.

ENTREPRENEURSHIP IN A CHANGING WORLD

Steve Burrill: "Whether we like to admit it or not, the United States is in the process of moving into a second tier. . . . We're moving to where we are no longer the dominant economic engine of the world."

The world is a very different place today than it has been for the past several decades, said G. Steven Burrill, CEO of the venture capital firm Burrill & Company. Burrill has played a key role in the growth of the U.S.

biotechnology industry over the past several decades, first with Ernst & Young, and more recently as a venture capitalist.

First, the marketplace has become flat, interconnected, and borderless. Ten years ago, pharmaceutical sales were predominantly in the United States, Europe, and Japan. By 2020, emerging markets are predicted to account for over half the growth in global sales.[2] "If you think about the dominant players in the world today, they are not well-empowered in the markets that are truly important," Burrill said.

Second, a powerful and growing middle class has emerged in countries that used to be largely poor. "The first thing that happens when people come out of a lower class poverty-driven society is they want health care. And they think health care is a right, not a privilege, so the demands on the health care system are enormous."

Third, new countries are coming to the fore, including the BRIC countries (Brazil, Russia, India, and China, to which Burrill added South Africa) and the CIVET countries (Columbia, Indonesia, Vietnam, Egypt, and Turkey). "Whether we like to admit it or not, the United States is in the process of moving into a second tier," Burrill added. "It's like Britain, a country with a proud past but kind of a lousy economic future. . . . We're moving to where we are no longer the dominant economic engine of the world."

Finally, every country is trying to build its economy around technology, and particularly biotechnology. "We need to be aware of that in the context of our competitive state today. We're no longer alone in what we're trying to do. Everybody's trying to do it around us."

Global Problems

The five biggest problems facing the world today, according to Burrill, are:

- Global climate change and the sustainability of the planet
- Clean water
- Energy security and self-sufficiency
- Food security and production
- Health care and health care reform

In Burrill's view, the life sciences can contribute to solving all of these problems. He used health care as an example of the changes going on today. The world is growing older. For example, China has more than 110 million

[2] Forecast by IMS Health, as reported in Reuters, *Emerging market drug sales seen $400 bln by 2020*, February 8, 2008, available at *uk.reuters.com/article/2008/02/27/uk-pharmaceuticals-emerging-idUKL2777827620080227.*

people over the age of 65.[3] This aging of the world population is creating tremendous demands on the health care system.

In addition, Burrill sees the focus of health care today moving toward chronic care as increasing progress is made against the diseases requiring acute care. As a result, the health care system needs to reorient care from hospital-based care toward wellness care where people will be sustained over long periods of time. In the future, people will be treated personally, predictably, and preemptively, which will dramatically change the opportunities for entrepreneurs. Health care will move from a cost-based system to a value-based system in which it looks for opportunities to add quality and value to life.

The integration of software and systems will drive wellness care. Rather than going to a hospital to receive care, people will wear a "smart t-shirt" with embedded sensors that will do all the monitoring that can be done in an intensive-care unit. Patients will feed the results of biological sample analyses into their cell phones before consulting with health care providers. Such a world will give rise to an entirely new ecosystem of entrepreneurs focused on the needs of the future, said Burrill.

Implications for Venture Capitalists

Despite the existence of technology clusters, new technologies can come from anywhere, said Burrill. Madison, Wisconsin, for example, is a center of biotechnology research and could become of a hotbed of commercial applications of biotechnology.

Furthermore, capital to develop new technologies is available everywhere, not just in the established centers of technology development. Even if venture capital is less available in the Midwest than elsewhere, the region has strong angel investor networks, Burrill observed. He also emphasized that venture capitalists may look at 100 to 200 deals a month and decide to support only one. "Most entrepreneurs talk to three VCs and say, 'Well, the VCs aren't interested.' Maybe you have to talk to 100 of these guys to find one. The tenacity necessary to ultimately find someone who's going to finance your company and build it is very important."

Entrepreneurs need to be willing to fail. "In Silicon Valley, we expect everybody to fail at least one or two times before they build a successful company."

Innovation has created extraordinary value, Burrill concluded. When Genentech, which Burrill helped develop, was sold to Roche, it was valued at $100 billion, which was more at the time than Pfizer, the biggest drug

[3] Central Intelligence Agency, The World Fact Book 2011, see *www.cia.gov/library/publications/the-world-factbook/geos/ch.html.*

company in the world. "We created more value in 25 years than did some of the biggest companies in the world." This innovation can occur anywhere in the world, so long as those regions are connected to the outside resources that make innovation possible. But people and companies need to allocate time to innovation. "You can't just hope it happens in your spare time. You have to make it a focus."

ATTRACTING REGIONAL INVESTMENTS

The Ohio Third Frontier Program, which was created in 2002, "supports applied research and commercialization, entrepreneurial assistance, early-stage capital formation, and expansion of a skilled talent pool that can support technology-based economic growth." It has been funded from a variety of sources, including the state budget, tobacco settlement money, and bonds approved by the voters. In May 2010, despite the recession, Ohio voters overwhelmingly approved a $700 million extension of the program.[4]

Frank Samuel, Jr., President of Geauga Growth Partnership, Inc., an economic development organization just outside Cleveland, pointed to several lessons that can be drawn from the program's success. First, bipartisanship has been essential for the program to be continued over time. Ohio governors and legislators from both parties have supported the program, partly because from the beginning it has been described and has operated in a bipartisan fashion.

Frank Samuel, Jr.: "If you can get financial returns, you will get jobs and economic growth and all the social benefits that come with it."

The second important lesson concerns the importance of having proposals for the program independently evaluated. The process has been intensely competitive, but potential grantees have known that grants were not being made on the basis of geographic fairness or political equity.

The third lesson is the importance of persistence. "Persistence without bipartisanship or independent evaluation would have been probably useless, but in our case it proved exceptionally valuable."

[4] Additional information on the Ohio Third Frontier Program is available at *www.thirdfrontier.com/ThirdFrontierCalendar/Default.aspx*.

The Great Lakes Economic Initiative

Samuel also described his involvement in the Great Lakes Economic Initiative, which was designed to establish a sustainable venture capital strategy for the Great Lakes region. Before the initiative, deals were too expensive in the region, said Samuel. As a result, companies based on discoveries made in the area took shape elsewhere.

Based on interviews with people in the region conducted under the auspices of the Brookings Institution, Samuel concluded that a \$1 billion to \$2 billion family of funds could achieve several needed ends. It could invest in the existing infrastructure of early stage venture funds. It would also co-invest in the most successful companies in the portfolios of those funds. And it could co-invest with a major fund from outside the region to create a significant presence in the region.

This approach would strengthen the availability of venture funds in every state in this region, said Samuel. It also would attract the kind of management needed to make ideas work. The approach is premised on financial returns, not on job creation or economic development. But job creation and economic development would be the result. "If you can get financial returns, you will get jobs and economic growth and all the social benefits that come with it."

An essential complement to the family of funds in the region is a support network of universities, technology transfer offices, research institutes, philanthropies, state and local programs, and other entities that can support economic growth and entrepreneurship. "They add value," Samuel said. "They are sufficiently searching in their criteria that the only projects that get across their thresholds are really good and merit investment by people who are making financial investments."

The financial community is not constrained by geography. If the financial community believes that there is money to be made in the Midwest, it will invest there. The challenge, said Samuel, is to figure out how to position the Midwest as a single community, because then it will be "extremely attractive to financial investors."

PERSPECTIVE FROM THE STATEHOUSE

Tommy Thompson, former governor of the State of Wisconsin and secretary of the U.S. Department of Health and Human Services, delivered a keynote speech at the workshop. He drew on his policy-making experience to make several observations about the goals of the workshop, particularly in the areas of education and economic development.

Thompson reviewed the recommendations of the original *Gathering Storm* report, and observed that the troubling fiscal condition of the federal government in coming years will make it difficult to increase spending on the K-12 education and research priorities the report identified. He provided perspectives on the challenges the United States faces in reigning in spending, on health care and the Medicare program in particular. The difficulty of addressing these challenges is compounded by increased polarization and gridlock in Congress. In Thompson's view, the increased ability of political parties to create legislative districts during redistricting that reliably vote for one or the other party is contributing to this polarization and gridlock.

In the education area, Thompson touched on the contribution that immigrant scientists and engineers make to U.S. innovation, an issue that was emphasized in the original *Gathering Storm* report. He pointed to the potential value of new approaches to immigration policy and education that would allow and encourage talented foreign-born scientists and engineers to stay in the United States after graduation and contribute to technology-based innovation and the broader economy.

He observed that during his time as governor of Wisconsin, state government was able to work with higher education, industry, and philanthropy to launch new university-based research programs that benefited the state and local economies. Although such collaboration is more difficult in the current environment, the Wisconsin Institutes of Discovery itself, built through the contributions of the state of Wisconsin, the Wisconsin Alumni Research Foundation, and John and Tashia Morgridge, stands as an encouraging example of what collaboration can accomplish.

Despite the serious challenges faced by the United States and by individual states and regions in sustaining an innovative 21st-century economy, Thompson expressed optimism that Americans will be able to come together and do what is necessary. He encouraged workshop participants to continue to their efforts to advance the *Gathering Storm* goals of strengthening research, education and innovation.

6

Final Observations

During the final session of the workshop, the participants engaged in a free-wheeling discussion of the important points they heard during the previous two days and steps to be taken next. Their individual observations and suggestions, which have been organized according to the four major sessions of the workshop, should not be taken as a consensus of the workshop participants as a whole or of the planning committee.

REVITALIZING K-12 SCIENCE AND MATH EDUCATION

- A coherent vision of the knowledge and skills that education should provide to students can drive improvement.
- Future generations will learn in different ways than have people in the past, which will require new and innovative approaches to education.
- Metrics for educational achievement among students and teachers can guide educational improvement.
- The informal STEM learning that occurs in such places as museums can have a powerful effect on both knowledge and attitudes.
- An emphasis on results rather than just funding can increase the interest of industry in contributing to K-12 education.
- Teachers need to be more adequately represented in discussions of education, perhaps through electronic connections from the schools where they are working.

- Giving teachers continuing education credits for learning how to apply for and manage grants could enable them to foster partnerships with the private sector.
- Teachers also need to spend time to learn through collaborative lesson planning and professional development.
- The valley of death plagues education as well, because few organizations exist that can develop promising innovations to the point that they can make a sustainable difference in the classroom.

STRENGTHENING UNDERGRADUATE SCIENCE AND ENGINEERING EDUCATION

- Revisions of undergraduate curricula across departments could produce a better alignment of undergraduate STEM education and workforce needs.
- Different states and regions have different needs that could be reflected in undergraduate STEM education.
- Two-year colleges are a critically important component of the higher education system in the United States.
- Recognition and support of students who are skilled at bringing others together and fostering achievement could produce major educational dividends.
- If more people could experience science as a means of exploring the unknown, they would better understand the process of bringing discoveries to the market to create jobs and wealth.

BUILDING EFFECTIVE PARTNERSHIPS

- Precompetitive cooperation in building production capacity can prepare the infrastructure needed for future production.
- Representatives from industry, K-12 education, and higher education rarely meet together, yet, as demonstrated by the conference, such meetings can be highly productive.
- Cooperation among academia, governments, and industry must be based on trust and on an appreciation of the value that each partner brings to the table.
- Partnerships succeed when all members of the partnership believe it to be to their advantage to make the collaboration work.
- Sharing information can build the trust necessary for collaborations to succeed.
- The Experimental Program to Stimulate Competitive Research (EPSCoR), an initiative by the National Science Foundation and other federal agencies to help build the research bases of jurisdic-

tions that have historically received relatively low levels of federal research funding, provides a valuable model of collaboration to achieve shared goals.

- Many barriers prevent faculty members from moving between academia and the private sector, despite the importance of such exchanges.
- Cooperation among a group of states that share common interests can yield better outcomes than competition.

FOSTERING REGIONAL TECHNOLOGY DEVELOPMENT AND ENTREPRENEURSHIP

- Innovation applies not just to products but to new approaches to technology development and entrepreneurship, such as new financing models.
- Financing for early-stage prototypes can demonstrate the feasibility of a product so that private industry will invest in its commercialization.
- A simple message is needed to convey the importance of science, innovation, and entrepreneurship to regional economies.
- Legislators tend to respond more positively to suggested solutions to problems than they do to requests for funding.
- Fellowship programs for scientists within state and local governments can create a connection between science and policy that is often missing.

Finally, Julie Underwood, the dean of the School of Education at the University of Wisconsin-Madison, observed that the conversation begun at the conference needs to be continuous, not a one-time event. "This conversation needs to go on and on."

Appendix A

Workshop Agenda

Rising Above the Gathering Storm
Developing Regional Innovation Environments

Wisconsin Institutes for Discovery
Madison, Wisconsin
September 20-22, 2011

Tuesday, September 20

Arrival and Opening Session

Noon	Registration opens
1:00 pm	Tours of Wisconsin Institutes for Discovery (on the hour)
5:30	Reception
6:30	Dinner
6:50	Welcome from Workshop Chair / **Judith Kimble,** Henry Vilas Professor and HHMI Investigator, University of Wisconsin
6:55	Welcome to the Wisconsin Institutes for Discovery / **Carl Gulbrandsen,** Managing Director, Wisconsin Alumni Research Foundation
7:00	Welcoming Remarks / **John P. Morgridge** and **Tashia F. Morgridge,** Founding Trustees, Morgridge Institute for Research
7:10	**Evening keynote: Creating a Regional Innovation Environment / Duane J. Roth,** CEO, CONNECT
7:45	Discussion
8:30	Adjourn

Wednesday, September 21

Regional Innovation Environments: Key Elements and Examples

8:30 am	Welcome/Introduction: Workshop Challenges and Goals / **Judith Kimble**
8:35	Setting the Stage: Five Years of Rising Above the Gathering Storm / **C. D. (Dan) Mote, Jr.**, Regents Professor and former President, University of Maryland
8:45	**Plenary Session One: Revitalizing K-12 Science and Mathematics Education.** Session chair, **Tom Luce**
8:45	The Foundation of Innovation: K-12 STEM Education / **Tom Luce**, CEO, National Math and Science Initiative
9:00	Ensuring Quality: A New Framework for Science Education Standards / **Helen R. Quinn**, Professor Emerita, SLAC
9:15	Delivering Value: Why K-12 STEM Education Matters in the Current Economy / **Michael Lach**, Special Assistant for STEM Education, U.S. Department of Education
9:30	Discussion on Revitalizing K-12 STEM Education
10:00	BREAK
10:30	**Plenary Session Two: Strengthening Undergraduate Education in Science and Engineering.** Session chair, **Lorrie A. Shepard,** Distinguished Dean of Education
10:30	Why Changing How We Teach Introductory Undergraduate Science Courses Is Critical for Our Nation's Future / **Bruce Alberts**, Editor-in-Chief, *Science*, and Professor Emeritus, Department of Biochemistry and Biophysics, University of California, San Francisco
10:45	Innovation in Teaching Undergraduate Biology / **Robin Wright**, Associate Dean, Department of Genetics, Cell Biology and Development, University of Minnesota
11:00	Improving Undergraduate Science and Engineering Education / **Lorrie A. Shepard,** Professor and Distinguished Dean of Education, University of Colorado, Boulder

11:15	Discussion on Strengthening Undergraduate STEM Education
11:45	BREAK
Noon	Working lunch: How can education help regional innovation environments thrive? What are the key strategies for implementation?
1:30 pm	**Plenary Session Three: Building Effective Partnerships among Governments, Universities, Companies, and Other Stakeholders for Innovation Environments.** Session chair, **Arun Majumdar**
1:30	The ARPA-E Model and the Innovation Ecosystem / **Arun Majumdar**, Director, ARPA-E
1:45	Government-University-Industry Partnerships: Global Innovation / **C. D. (Dan) Mote, Jr.**, Regents Professor and Former President, University of Maryland
2:00	New Alliances among Government, Industry and Universities / **Mary Good**, Donaghey University Professor, University of Arkansas
2:15	Discussion on Building Effective Partnerships
2:45	BREAK
3:15	**Plenary Session Four: Fostering Regional Technology Development and Entrepreneurship.** Session chair, **William J. Spencer**, Chairman Emeritus, SEMATECH
3:15	Innovative Research Environments for Regional Development / **Sangtae Kim**, Executive Director, Morgridge Institute for Research
3:30	Fostering Entrepreneurship and Venture Investment in Regions / **Frank Samuel, Jr.**, President, Geauga Growth Partnership, Inc
3:45	Venture Investment and Regional Development in the Life Sciences Sector / **G. Steven Burrill**, CEO, Burrill & Company
4:00	Discussion on Fostering Regional Technology and Entrepreneurship
4:30	BREAK

5:30	Reception
6:00	**Keynote speaker: Honorable Tommy Thompson, Former Governor of Wisconsin and Secretary of the U.S. Department of Health and Human Services**
6:45 pm	Discussion
7:00	Working dinner: How can partnerships and entrepreneurship help regional innovation environments thrive? What are the key strategies for implementation?

Thursday, September 22

Summing Up and Next Steps

8:00 am	Breakfast: (Panelists and table leaders to be organized privately)
9:00	**Panel discussion: Mary Good, Duane Roth, Helen Quinn**
	Discussion of Key Themes and Strategies for Implementation

11:00 Adjourn

Appendix B

Biographical Sketches of Agenda Speakers and Planning Committee Members

Bruce Alberts (Plenary Session Two Speaker, Planning Committee Member) is a prominent biochemist with a strong commitment to the improvement of science and mathematics education. He is editor-in-chief of *Science* and professor emeritus of biochemistry and biophysics at the University of California, San Francisco, to which he returned after serving two six-year terms as the president of the National Academy of Sciences. He also serves as one of President Obama's first science envoys. He received his A.B. and Ph.D. degrees from Harvard University.

G. Steven Burrill (Plenary Session Four Speaker) is chief executive officer of Burrill & Company. He has been involved in the growth and prosperity of the biotechnology industry for over 40 years. Prior to founding Burrill & Company in 1994, he spent 28 years with Ernst & Young, directing and coordinating the firm's services to clients in the biotechnology/life sciences/high technology/manufacturing industries worldwide. Mr. Burrill holds a B.A. degree from the University of Wisconsin.

Ruth A. David (Planning Committee Member) is president and CEO of ANSER (Analytic Services Inc.), a not-for-profit corporation that provides research and analytic support on national and transnational issues. She was previously deputy director for Science and Technology at the Central Intelligence Agency. A member of the National Academy of Engineering, Dr. David received a B.S. from Wichita State University, and M.S. and Ph.D. degrees in electrical engineering from Stanford University.

Mary Good (Plenary Session Three Speaker, Final Panel) is the Donaghey University Professor at the University of Arkansas, Little Rock and serves as the managing member for Venture Capital Investors, LLC. Dr. Good was previously U.S. under secretary of Commerce for Technology and senior vice-president of technology at Allied Signal, Inc. A member of the National Academy of Engineering, she received a B.S. from the University of Central Arkansas and M.S. and Ph.D. degrees from the University of Arkansas.

Carl E. Gulbrandsen (Welcome) is the managing director of Wisconsin Alumni Research Foundation, the patent management organization for the University of Wisconsin Madison. He joined WARF in 1997, after practicing intellectual property law in private practice and with several high technology companies. He received his B.A. degree from St. Olaf College, and his Ph.D. in physiology and J.D. from the University of Wisconsin-Madison.

Sangtae Kim (Plenary Session Four Speaker) is executive director of the Morgridge Institute for Research. Located on the University of Wisconsin-Madison campus, the institute is intended to become the Midwest's premier, private medical research institute. Prior to his appointment at the Morgridge Institute, Dr. Kim served on the faculties at Purdue University and the University of Wisconsin-Madison, and also held positions in government and industry. A member of the National Academy of Engineering, Dr. Kim earned B.S. and M.S. degrees from the California Institute of Technology, and a Ph.D. in chemical and biological engineering from Princeton University.

Judith Kimble (Workshop and Planning Committee Chair) is Henry Vilas Professor at the University of Wisconsin-Madison in the Departments of Biochemistry and Medical Genetics and an investigator with the Howard Hughes Medical Institute (HHMI). Her research focuses on the molecular regulation of animal development. Over the course of her career, she has made seminal contributions in the area of how stem cells are regulated to self-renew or differentiate. A member of the National Academy of Sciences, Dr. Kimble earned her B.A. at the University of California-Berkeley and her Ph.D. at the University of Colorado-Boulder.

Michael Lach (Plenary Session One Speaker) is special assistant for STEM Education, U.S. Department of Education. Previously, he was officer of teaching and learning for Chicago Public Schools. Mr. Lach began his professional career teaching high school biology and general science in New Orleans in 1990 as a charter member of Teach for America. He earned a bachelor's degree in physics from Carleton College, and master's degrees from Columbia University and Northeastern Illinois University.

Tom Luce (Plenary Session One Chair and Speaker) recently stepped down as chief executive officer of the National Math and Science Initiative, which was created in 2007 to implement the recommendations of *Rising Above the Gathering Storm* by dramatically improving U.S. K-12 math and science education. He previously served as U.S. assistant secretary of Education for Planning, Evaluation and Policy Development. An attorney, Mr. Luce received his undergraduate and graduate degrees from Southern Methodist University.

Arun Majumdar (Plenary Session Three Chair and Speaker) became the first director of the Advanced Research Projects Agency-Energy (ARPA-E) in October 2009. Dr. Majumdar was previously associate laboratory director at Lawrence Berkeley National Laboratory and on the faculty of the University of California, Berkeley. A member of the National Academy of Engineering, he received his B.S. at the Indian Institute of Technology, Bombay and his Ph.D. from the University of California, Berkeley.

John P. Morgridge (Welcome) is a founding trustee of the Morgridge Institute for Research, a trustee of the Wisconsin Alumni Research Foundation, and chairman emeritus of Cisco Systems. He was president/CEO of Cisco from 1988 to 1995, growing the company from $5 million to over $1 billion in sales. He previously held top management positions at several other information technology companies, and served in the U.S. Air Force. He received a BBA from the University of Wisconsin-Madison and an M.B.A. from Stanford University.

Tashia F. Morgridge (Welcome) is a founding trustee of the Morgridge Institute for Research and a member of the University of Wisconsin, School of Education Board of Visitors. She and her husband John actively support a range of education, conservation, and human services initiatives, including generous support for the Morgridge Institute for Research and numerous other initiatives at the University of Wisconsin. Mrs. Morgridge was previously a special education teacher. She received a B.S.E. from the University of Wisconsin-Madison and an M.S. from Lesley College.

C.D. (Dan) Mote, Jr. (Setting the Stage, Plenary Session Three Speaker, Planning Committee Member) is Regents professor and Glenn L. Martin Institute Professor of Engineering at the University of Maryland. He served as president of the University of Maryland from September 1998 to August 2010, spurring the university to lead the state in the development of its high technology economy. A member of the National Academy of Engineering, Dr. Mote received B.S., M.S., and Ph.D. degrees in engineering, mechanics from the University of California at Berkeley.

Paul S. Peercy (Discussion Leader) is dean of the College of Engineering at the University of Wisconsin-Madison. He was previously president of SEMI/SEMATECH, a non-profit technical R&D consortium of U.S.-owned and operated companies that comprise the equipment and supplier infrastructure for the semiconductor industry. A member of the National Academy of Engineering, Dr. Peercy earned a B.S. from Berea College, and M.S. and Ph.D. degrees in physics from the University of Wisconsin.

Helen R. Quinn (Plenary Session One Speaker, Final Panel) is a professor of physics emerita at Stanford University where she also served as Education and public outreach manager at the SLAC National Accelerator Laboratory. She chairs the National Research Council's Board on Science Education, and chaired the 2011 study *A Framework for K-12 Science Education: Practices, Crosscutting Concepts, and Core Ideas.* A theoretical physicist and National Academy of Sciences member, Dr. Quinn received her B.S., M.S., and Ph.D. degrees in physics from Stanford University.

Duane J. Roth (First Evening Keynote, Final Panel) is chief executive officer and member of the board of CONNECT, a nonprofit organization dedicated to creating and sustaining the growth of innovative technology and life science businesses in San Diego. Prior to joining CONNECT, he founded Alliance Pharmaceutical Corp., where he serves as chairman of the board, and held senior management positions at Johnson & Johnson and American Home Products (now Pfizer) operating companies. Mr. Roth earned a B.S. from Iowa Wesleyan College.

Frank Samuel, Jr. (Plenary Session Four Speaker) is president of the Geauga Growth Partnership, Inc., a business-led economic development organization in Geauga County (northeastern Ohio). He was science and technology advisor to the governor of Ohio from 2000–2007, where he was a principal architect of Ohio's Third Frontier Project. Mr. Samuel is a graduate of Hiram College and Harvard Law School.

Lorrie A. Shepard (Plenary Session Two Chair and Speaker) is University Distinguished Professor and dean of the School of Education at the University of Colorado at Boulder. Her research focuses on psychometrics and the use and misuse of tests in educational settings. She was elected to the National Academy of Education in 1992 and served as its president from 2005 to 2009. She earned her B.A. from Pomona College, and her M.A. and Ph.D. from the University of Colorado at Boulder.

William J. Spencer (Plenary Session Four Chair, Planning Committee Member) is chairman emeritus of International SEMATECH, having served as chairman of the SEMATECH and International SEMATECH boards, and previously as SEMATECH's president and chief executive officer. He also held key research and management positions at Xerox Corporation, Bell Laboratories, and Sandia National Laboratories. A member of the National Academy of Engineering, Dr. Spencer received an A.B. degree from William Jewell College an M.S. degree in mathematics and a Ph.D. in physics from Kansas State University.

The Honorable **Tommy G. Thompson** (Keynote Speaker) is currently a partner at Akin Gump Strauss Hauer & Feld LLP. Before entering the private sector in 2005, Secretary Thompson enjoyed a long and distinguished career in public service, including service as U.S. secretary of Health and Human Services and 14 years as governor of Wisconsin. Secretary Thompson received both his B.S. and J. D. from the University of Wisconsin-Madison.

Julie Underwood (Final Panel) is dean of the School of Education at the University of Wisconsin-Madison. She was previously dean of Miami University's School of Education and Allied Professions. Dr. Underwood has a bachelor's degree in political science and sociology from DePauw University, a law degree from Indiana University, and a Ph.D. in educational leadership from the University of Florida.

Robin Wright (Plenary Session Two Speaker) is associate dean for Faculty and Academic Affairs in the College of Biological Sciences (CBS) and professor of Genetics, Cell Biology, and Development at the University of Minnesota. Her major goal as associate dean is to catalyze the development of the nation's best biology curriculum. Her research focuses on the genetics and physiology of cold adaptation in yeast. Dr. Wright earned a B.S. degree from the University of Georgia and a Ph.D. from Carnegie-Mellon University.

Appendix C

Workshop Participant Roster

Paul Ahlquist
Morgridge Institute for Research/
HHMI

Bruce Alberts
Science Magazine

Kristine Andrews
UW System Administration

Tony Armstrong
Indiana University

Tom Arrison
The National Academies

Lauren Baker
Milwaukee Public Schools

Linda Bartlett
New Era

Jay Bayne
Milwaukee Institute

Jane Belmore
Edgewood College

Ellen Bergfeld
American Society of Agronomy

Anthony Boccanfuso
University-Industry Demonstration
Partnership

Eric Brunsell
University of Wisconsin-Oshkosh

G. Steven Burrill
Burrill & Company

John Burris
Morgridge Institute for Research

Edward Clarke
Madison Area Technical College

Harvey Cohen
Morgridge Institute for Research

Marta Collier
Arkansas Science & Technology Authority

Kevin Conroy
Exact Sciences Corporation

Kathe Crowley Conn
Aldo Leopold Nature Center

Maria Dahlberg
Pennsylvania State University

James Dahlberg
Morgridge Institute for Research

Mary Darrow
Iowa State University

Troy Dassler
Wisconsin Alumni Research Foundation

Lorelei Davis
Michigan State University

Ian Davison
Central Michigan University

Paul DeLuca, Jr.
University of Wisconsin

Tim Donohue
University of Wisconsin

Michael Ferris
University of Wisconsin

Kevin Finneran
The National Academies

Paul Fowler
Wisconsin Institute for Sustainable Technology

Adam Gamoran
University of Wisconsin

Richard George
Great Lakes Higher Education Corp.

Mary Good
University of Arkansas-Little Rock

Peter Goodwin
University of Idaho

J. Ian Gray
Michigan State University

Carl Gulbrandsen
Morgridge Institute for Research

Mary Gulbrandsen
Fund for Wisconsin Scholars

Jennifer Harper-Taylor
Siemens Foundation

Charles Hasemann
Michigan State University

Laura Heisler
Wisconsin Alumni Research Foundation

Katy Heyning
School of Education
University of Wisconsin-Whitewater

Richard Hichwa
University of Iowa

Robert Hoar
University of Wisconsin-La Crosse

Theodore Imes
Northrop Grumman Electronic Systems

Jorge Jose
Indiana University

Michael Kaspar
National Education Association

Terry Kelly
Aldo Leopold Nature Center

Aaron Kershner
University of Wisconsin

Ryan Kershner
Thermo Fisher

Kurt Kiefer
Wisconsin Department of Public Instruction

Sangtae Kim
Morgridge Institute for Research

Judith Kimble
University of Wisconsin

Betsy Kippers
Wisconsin Education Association Council

Michael Lach
U.S. Department of Education

Pat Lipton
Morgridge Institute for Research

Miron Livny
Morgridge Institute for Research

Tom Luce
National Math & Science Initiative

Arun Majumdar
ARPA-E

Ernie Micek
Morgridge Institute for Research

Susan Millar
Wisconsin Institutes for Discovery

Gary Mirka
Iowa State University

John Morgridge
Morgridge Institute for Research

Tashia Morgridge
Morgridge Institute for Research

C. D. (Dan) Mote, Jr.
University of Maryland

Daniel Nerad
Madison Metropolitan School District

Peter Nordgren
University of Wisconsin-Superior

Rhonda Norsetter
Chancellor's Office
University of Wisconsin

Paul Peercy
University of Wisconsin

Terry Potter
Morgridge Institute for Research

Helen Quinn
Board on Science Education
National Research Council

Fred Robertson
Morgridge Institute for Research

Roy Romer
The College Board

Duane Roth
CONNECT

Bob Rothschild
Morgridge Institute for Research

Sara Rothschild
Morgridge Institute for Research

Carmel Ruffolo
University of Wisconsin-Milwaukee
University of Wisconsin-Parkside

Steven Salter
Project Lead the Way-Wisconsin

Frank Samuel, Jr.
Geauga Growth Partnership

Gary Sandefur
University of Wisconsin-Madison

Lorrie Shepard
University of Colorado at Boulder

Rupa Shevde
Morgridge Institute for Research

Tatsuya Shinkawa
NEDO

William J. Spencer
SEMATECH

Tom Spitz
Settlers Bank

Becky Splitt
Study Blue Inc.

Tommy G. Thompson
Akin Gump Strauss Hauer & Feld LLP

Lorin Toepper
Madison Area Technical College

Julie Underwood
School of Education
University of Wisconsin

William Vajda
City of Marquette, Michigan

David Ward
University of Wisconsin

Robin Wright
University of Minnesota

Tom Zinnen
University of Wisconsin-Madison Extension